W9-BYY-044

YOUR ENVIRONMENT

BRENDA WILLIAMS

Discard
NHCPL

RAINTREE
STECK-VAUGHN
PUBLISHERS
The Steck-Vaughn Company

Austin, Texas

NEW HANOVER COUNTY
PUBLIC LIBRARY
201 CHESTNUT STREET
WILMINGTON, NC 28401

© Copyright 1999, text, Steck-Vaughn Company

All rights reserved. No part of this book may be
reproduced or utilized in any form or by any means,
electronic or mechanical, including photocopying,
recording, or by any information storage and retrieval
system, without permission in writing from the Publisher.
Inquiries should be addressed to: Copyright Permissions,
Steck-Vaughn Company, P.O. Box 26015, Austin, TX 78755.

Published by Raintree Steck-Vaughn Publishers,
an imprint of Steck-Vaughn Company

Library of Congress Cataloging-in-Publication Data
Williams, Brenda.
Your environment / Brenda Williams.
     p.   cm.—(Geography starts here)
     Includes bibliographical references and index.
     Summary: Introduces concepts about the environment,
     discussing the land, water, air, and recycling.
     ISBN 0-8172-5117-0
     1. Environmental sciences—Juvenile literature.
     2. Human ecology—Juvenile literature.
     [1. Environmental Protection. 2. Environmental sciences.]
     I. Title. II. Series.
     GE115.W55 1998
     363.7—dc21   97-43122

Printed in Italy. Bound in the United States.
1 2 3 4 5 6 7 8 9 0 03 02 01 00 99

Printed and bound by L.E.G.O. S.p.A., Vicenza, Italy
Picture Acknowledgments
Pages 1: Lionheart Books. 4: Wayland Picture Library. 5: Eye Ubiquitous/Michael George.
6: Zefa Photo Library. 8: Eye Ubiquitous/H. Rooney. 9: Eye Ubiquitous/A. Carroll. 10–11: Zefa
Photo Library. 11: Wayland Picture Library/Chris Fairclough. 12: Wayland/Axiom, Jim Holmes.
14: Aerofilms Ltd. 15: Zefa Photo Library/Weir. 16–17: Zefa Photo Library/Stockmarket.
18: Lionheart Books. 19: Eye Ubiquitous/L. Fordyce. 20–21: Wayland Picture Library. 22: Eye
Ubiquitous/J. Waterlow. 23: Zefa Photo Library/E. M. Bordis. 24: Eye Ubiquitous/John Hulme.
25: Zefa Photo Library/Stockmarket/J. B. O'Rourke. 26–27: Zefa/Stockmarket/John Sohm. 27:
Eye Ubiquitous/James Paul. 28: Zefa/Stockmarket/Gabe Palmer. 29: Eye Ubiquitous/David
Cumming. 30: Ecoscene/Robert Nichol. Cover: Zefa Photo Library/Streichan.

The photo on the title page shows the city of Marseilles, France.

CONTENTS

WHAT IS YOUR ENVIRONMENT?

Everything around you makes up your environment. Your environment is all you can see, feel, hear, and smell. It is all the things that are alive and those that are not. You are part of your environment, too!

There are many kinds of environments. Your room, home, garden, street, park, school, and town are all environments. So are forests, jungles, deserts, and oceans.

Farmland is a mixture of natural and artificial environments. This farm is in Egypt.

A view of New York City.
It wasn't always a city.
People can change their
environments.

THE LIVING WORLD

All plants and all animals (including humans) get all they need to live from the world around them. The air that we breathe, the food that we eat, and the water that we drink all come from our environment.

Wildlife rangers help children explore a woodland.

Some plants and animals provide food for us, as well as materials for clothing, energy, and shelter.

THE WEB OF LIFE

Plants and animals in your environment depend on one another.

1. Sunlight provides the energy plants need to make their own food.

2. We cannot make our own food. We get food by eating plants, or other animals, in our environment.

3. Plants and animals need warmth. Sunshine makes the environment suitable for life.

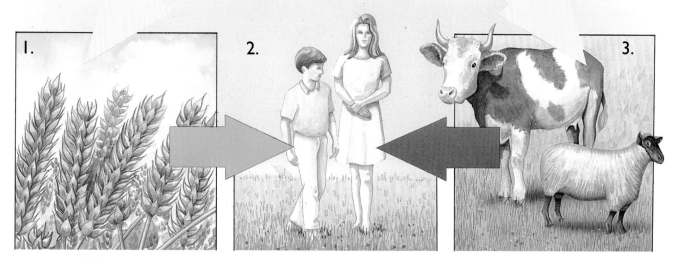

1.

2.

3.

A Changing World

Our environment changes all the time—
with the seasons and the weather. People
can change their environment, too.

Farmers change the way the countryside
looks by plowing fields, growing crops, and
draining water from their land.

This area of countryside is
being cleared to build a
road. We need good
roads, but we must not
damage our surroundings
to build them.

Villages grow into towns. Roads, bridges, and factories are built where there once were forests and meadows.

This builder uses lumber and bricks from old houses rather than new materials from the environment.

WORKING THE LAND

Enough children are born in the world every week to fill a new city. More and more land is used to build homes and to grow food for the increasing numbers of people.

To grow more food, farmers put fertilizers on their land. These kill weeds and insect pests with chemicals. But fertilizers and chemicals can also pollute soil, rivers, and animals.

This farmer in China uses cattle power to plow his land.

▼ Farmland, fields, and woods outside towns form "green belts" that keep the environment looking nice.

Settlements

Most people live in towns and cities. These environments have been made by people.

Towns use a lot of natural resources—water, gas, oil, and stone—for homes, schools, shops, offices, and factories.

Where there are settlements, there are usually cars. Fumes from this traffic in Seoul, South Korea, fill the air. Burning fuels such as gasoline creates gases that pollute, or dirty, the air.

THE HOME ENVIRONMENT

People, their homes, and cars use up natural resources and pollute the environment.

Chimneys let out smoke from fires.

Pipes and cables under the streets bring water, gas, and electricity to our homes.

Underground pipes carry away dirty water to nearby sewers and rivers.

Roads carry people
and goods to and
from factories.

Industrial Areas

Factories are busy, noisy places, with people and goods constantly coming and going. Factories are usually built far from homes.

Factories are often built near rivers, because they use a lot of water. Many of the goods we use—clothes, toys, cars, furniture—are made in factories. But factories also make waste, which may include harmful chemicals.

At this chemical factory in Scotland, large chimneys carry waste steam high into the air.

Natural Resources

Everything around us is made from natural resources, such as oil, lumber, rocks, and metals.

People have used these natural resources for thousands of years. They will not last forever.

Huge pits are dug in the ground to get building materials for houses, roads, and factories.

We can get energy from wind (wind power), sea (tidal power), and sunlight (solar energy). These are "renewable" resources. They will not run out, and they do not produce wastes that harm the environment.

The environment provides most of the building materials needed for houses, cars, and driveways.

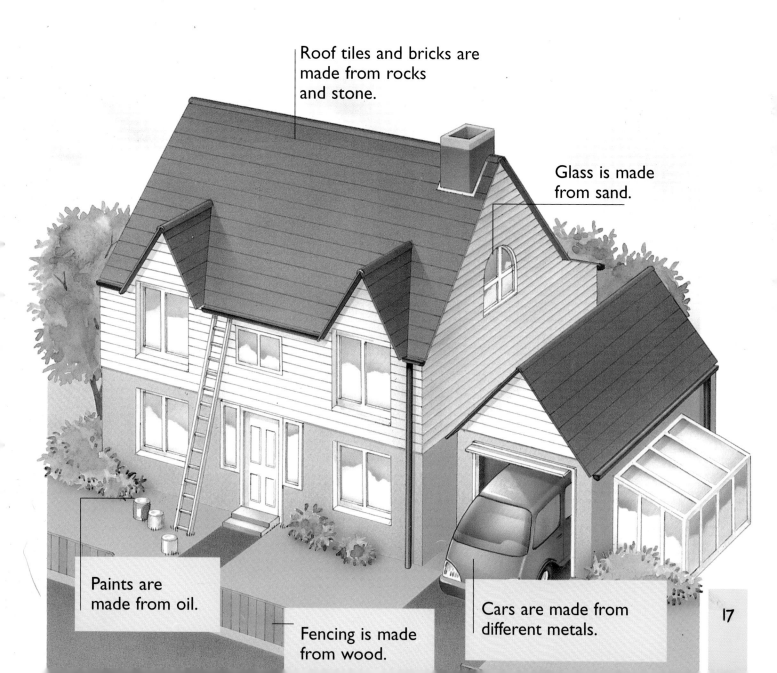

Roof tiles and bricks are made from rocks and stone.

Glass is made from sand.

Paints are made from oil.

Fencing is made from wood.

Cars are made from different metals.

17

THE WATERY WORLD

Vacations by the sea are fun. But vacationers disturb animals, such as turtles, that lay their eggs on quiet beaches.

We use water to drink, wash, and cook. Water powers some machines and puts out fires. We swim, sail, and surf on water. We fish in rivers, lakes, and seas for food.

We also pump waste from drains into rivers and seas. Oil from ships may spill out into the ocean. Yet plants and animals can only live in clean water.

Big fishing boats scoop up huge amounts of fish in their nets. The nets may also trap animals such as dolphins and turtles.

Using Rivers

People have always lived near rivers because they need fresh water to drink. Rivers make travel easy, too. Boats can carry goods and people where there are no roads.

Cleaning Water

Fill a jar with water from a stream or pond. Push a thick coffee filter into the neck of another jar. Pour some of the pond water into the filter. The paper holds back any dirt in the water. See the difference between the "clean" and "dirty" pond water.

Water to be filtered

Mud and sand trapped in the filter

Collecting jar

Clean water

Coffee filter

"Dirty" pond water

Bridges carry traffic over the Seine River, in France, and barges carry goods along it.

River water can also power machinery. Watermills used to grind corn. Today, hydroelectric stations are used to make electricity from fast-running water.

THE AIR WE BREATHE

All around and above us is air—a mixture of gases we cannot see or smell. Plants need gases from the air to make food. Animals need oxygen to breathe.

Air and temperature make our weather and change our environment. Winds blow clouds through the sky. Clouds carry water that falls as rain. Sunshine dries water from the ground and turns it back into clouds.

On windy days we feel the air moving. This hang glider uses wind to keep him up in the air.

The weather affects our environment—the clothes we wear, the food we grow, and the homes we build.

Smoke and Dust

When you have a bonfire or a barbecue, smoke rises into the air. Smoke also comes from factory chimneys. Mixed in this smoke are chemicals that can pollute the air.

Power plants, cars, and even herds of cows all give off gases into the air. In some cities, car exhausts make a hazy smog in which people find it difficult to breathe.

Saving Energy

We can burn less coal, oil, and gas if we stop wasting energy. Turn off lights when you leave a room. Do not use a car to make short journeys. Wrap up warmly in winter instead of turning up your furnace.

This traffic policeman in Bangkok, Thailand, is wearing a mask that filters out pollution in the air.

Smog hangs over the city of Los Angeles. Smog turns a clear sky hazy.

THINKING AHEAD

We can all care for our environment. Streets and parks can be made free from litter and graffiti. The air can be cleaned up by using fewer cars. We can repair shabby buildings.

Roads are jammed with traffic. Yet one train can carry hundreds of people and truckloads of goods. More people should choose to travel by train and bicycle. We need more bike paths and city centers free from traffic.

Trolleys, trains, and bicycles create less pollution and do not clog the roads with traffic.

▼ Biosphere 2, a greenhouse-like building in which scientists lived for months by growing their own plant food and recycling air and water

Recycle, Re-use, Repair

We throw away tons of garbage each day. But paper, metal cans, glass bottles, and old clothes can all be recycled or re-used.

Recycled paper is turned into toilet paper and newsprint. Metal and glass can be melted down and used again.

At a recycling factory, tin and aluminum cans are sorted as they pass along conveyor belts.

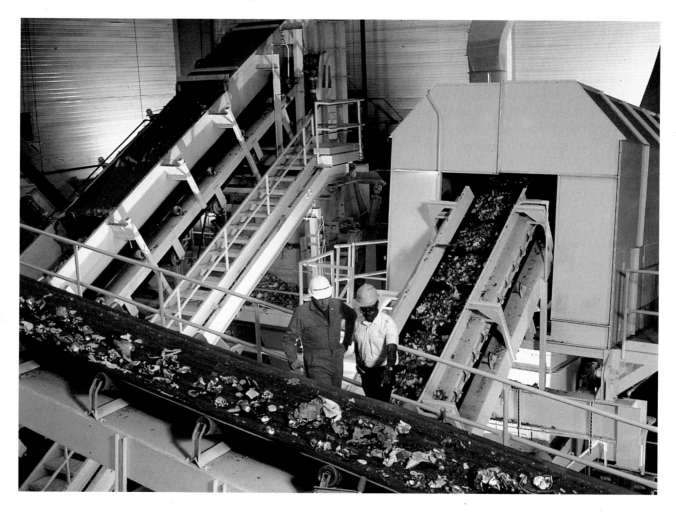

Recycling

Go with an adult to your local recycling center and see what it collects. By recycling items, such as paper, cardboard, glass, and clothes, we will use less of the earth's precious resources and create less pollution.

Wherever we live in the world, we share our environment with all other living things. We must take care of plants and animals so that we can survive. If we are careful, we can protect our environment for everyone.

People sort through garbage from offices in Delhi, India, to recycle paper, glass, and cardboard.

FACTS AND FIGURES

The first farmers

Until about 10,000 years ago the earth only had natural environments. This began to change when people started to farm the land.

The start of the end

In the 1700s, the Industrial Revolution began. New machines were invented, and factories were built. Raw materials, such as coal and iron, were used more than ever before to keep the machines working. These resources are fast running out.

The move to the city

In 1800 only one person in 20 lived in a city. Today about half the people in the world live in cities. Mexico City and Tokyo are the two biggest cities in the world. Each has a population of about 20 million people.

Machines are used to cut down trees in forests to get lumber for houses.

Water, water everywhere

Water covers two-thirds of the earth's surface and land a third. There is also water in the air—in clouds. Most of the earth's water is salty —as in the seas and oceans. Of the fresh water, two-thirds exists in the ice or snow of the Arctic and Antarctic regions. Only a third is found in rivers and lakes.

Population explosion

Since the 1600s the population of our planet has grown from about 450 million to nearly 6 billion.

Further Reading

Anderson, Joan. (photos by George Ancona). *Earth Keepers*. San Diego: Harcourt Brace, 1993.

Bailey, Donna. *What We Can Do About Protecting Nature* (What We Can Do). Danbury, CT: Franklin Watts, 1992.

Benson, Laura. *This Is Our Earth*. Watertown, MA: Charlesbridge Publishing, 1994.

Pollock, Steve. *The Atlas of Endangered Places*. New York: Facts on File, 1993.

Wheeler, Jill C. *The People We Live With* (We Can Save the Earth). Edina, MN: Abdo and Daughters, 1991.

GLOSSARY

Artificial Anything that is not natural and has been made by people.

Energy The power needed to make anything, including living things, work and move. All life depends on energy from the sun.

Factories Buildings in which goods are made or processed.

Fertilizers Chemicals that are used to help plants grow and give nutrients to soil.

Fuels Materials that we burn to produce energy.

Hydroelectric plants Places where electricity is made by using the flow of water from rivers or from lakes stored behind dams.

Litter Garbage that we throw away, such as fast-food cartons, drink cans, paper wrappers, and bottles.

Oxygen A gas found in air, breathed by animals and people.

Pollute To dirty the air, water, or land with waste.

Raw materials Resources from the environment, such as coal, oil, timber, and metals. Raw materials are used to make goods.

Recycling Using materials again and again, instead of throwing them away after one use.

Renewable resources Things such as sunshine, wind, and water that can make energy without being used up. These resources will not run out.

Resources Anything in the environment that we use, such as air, water, and minerals.

Smog Smoke or pollution-filled fog.

Trolleys Passenger vehicles that run on rails and are driven by electricity, not fuel.

INDEX

© Copyright 1997 Wayland (Publishers) Ltd.